Copyright © 2009 by James Lu Dunbar

All rights reserved. No part of this book may be reproduced in any form or by any electronic or mechanical means, without permission in writing from the publisher, except by a reviewer who may quote brief passages in a review.

Any members of educational institutions wishing to reproduce part or all of the work for non-commercial classroom use has the writer's and publisher's permission to do so.

James Lu Dunbar
2140 Shattuck Avenue #2406
Berkeley CA 94704

www.JLDunbar.com

BANG!

written & illustrated by James Lu Dunbar

BANG!

The Universe Verse Book 1

The universe is full of magical things,
patiently waiting for our wits to grow sharper.

- Eden Phillpots

In the beginning, before **Time** had begun, NOTHING existed, and nothing was fun.

to highly complex and **HUGELY ENORMOUS.**

Still growing to this day, with remarkable pace,

creating as it goes, the existence of place.

Since beyond our expanding universe face,

nothing exists,

not even SPACE!

At this point in the game, there's just one thing to name.

Everything's **Energy,**

all one and the same.

Everything ruled by a singular **FORCE,** so all of existence is from the same source.

As it cooled down further, there came the first splatter of what was to come, just a split second latter – a new form of energy: the first-ever **MATTER.**

The creation of matter was a quite frightful storm, as energy changed into physical form.

Each particle was born as part of a pair, so the creation of matter was balanced and fair.

If a particle met with its opposite double, back into energy, they'd pop like a bubble.

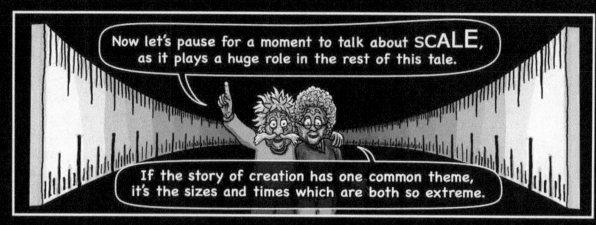

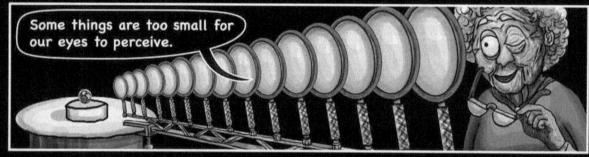
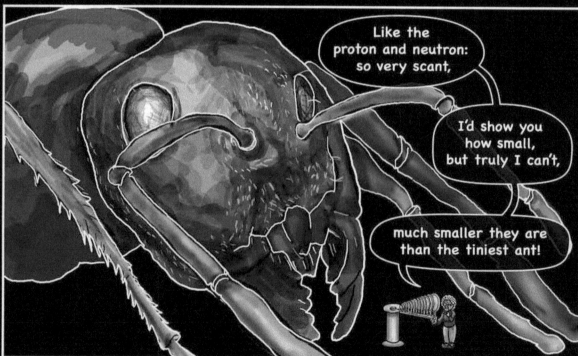

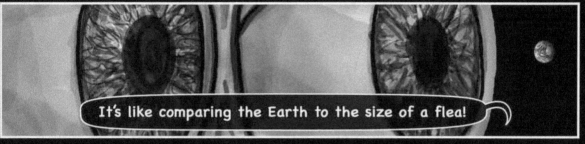

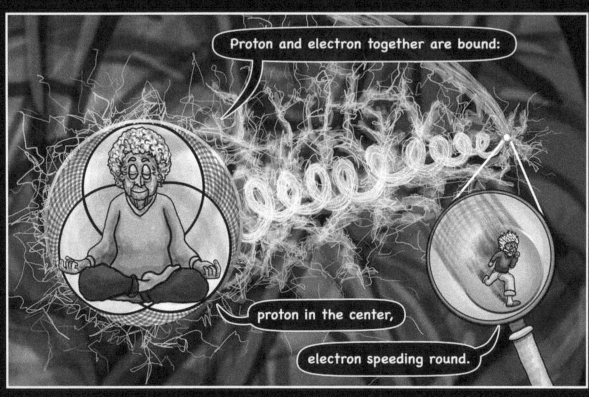

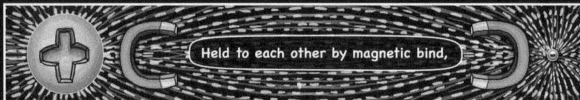

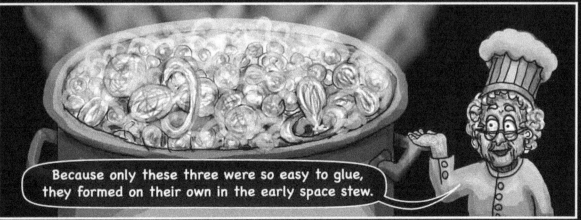

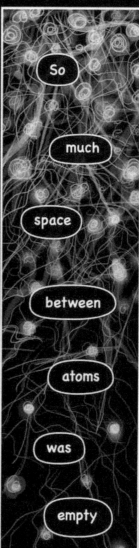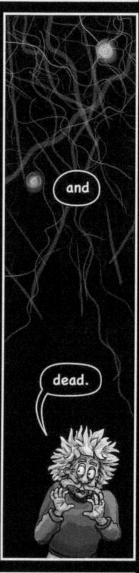

Gravity gives matter a small constant pull, great for a universe that wasn't so full.

So every last atom, through all of existence, is pulled to the others by gravity's insistence.

The strength of this force gets smaller with distance, so the power of its pull comes from persistence.

Over time, it acted as a great cosmic girdle:

The thinly spread matter did eventually curdle.

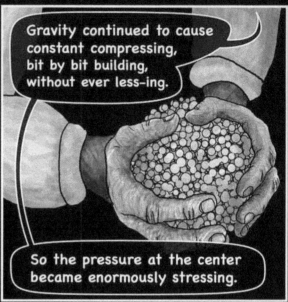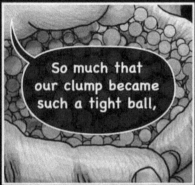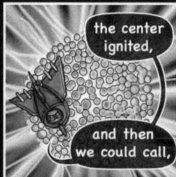

In this hotter than hot, most pressure-filled zone,

they are so close together

it becomes quite unknown, as the atoms forget

which electrons they own.

WARNING POSSIBLE FUSION

And if that wasn't enough, not knowing who's whose,

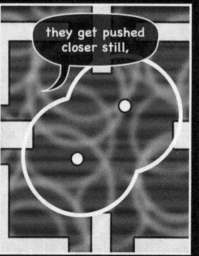

they get pushed closer still,

'till whole atoms fuse.

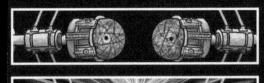

in accordance with that most famous equation...

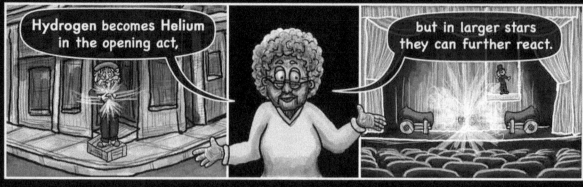

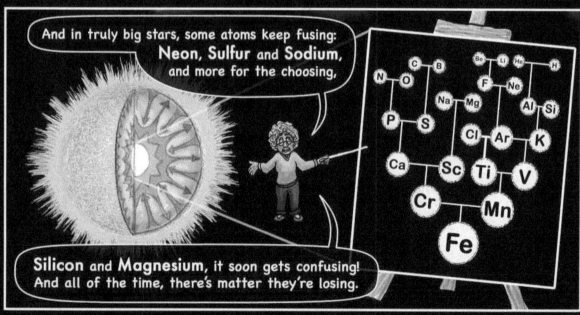

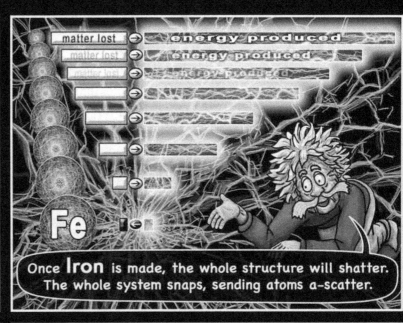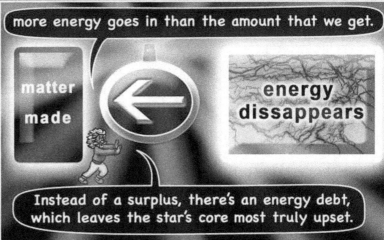

Now the star gets collapsed 'til it's dense as can be.

But here it turns out, the star has a Plan B.

Those atoms, squeezed tight, are left with no room,

to the point where the joining of atoms can resume.

This last flurry of fusion sends the whole thing Ka-BOOM!

As a **SUPERNOVA** explosion goes out with great zoom.

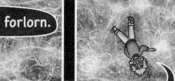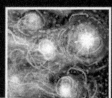

Millions of stars gathered into formations, of great spiral disks, with mighty rotations.

GALAXY

is the name for these giant star groups. Gravity holds them in spinning star loops.

BANG!

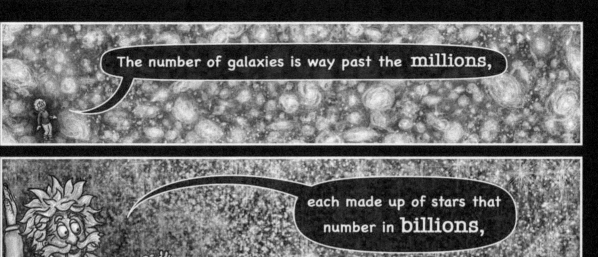

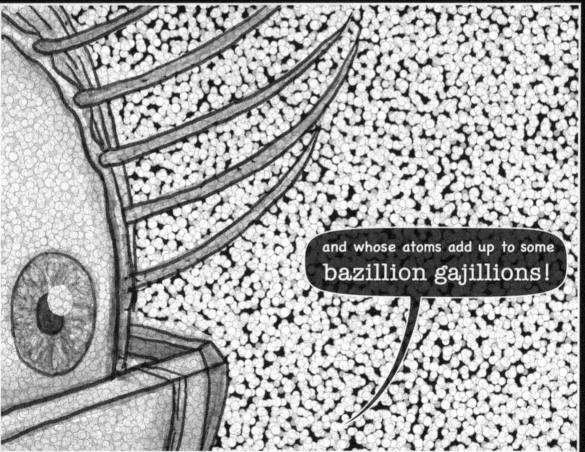

And that's how all that we know was created.

In a blink before time — a dot that dilated.

Energy into matter as the whole thing inflated.

Growing and growing, it has never abated.

Powered by forces that repel and attract,

particles of matter were impelled to interact.

In the hearts of the stars they were forced to react.

Atoms fused into one, through gravity's impact.

BANG!

These new types of atoms are a cause for delight,

as atomic variation is sure to invite,

new levels of structure

to **amaze**

and **excite!**

To be continued...

A Note from the Author:

I am not a professional scientist,
just an enthusiastic amateur with a library card.
I was able to write this book because of the countless people
who have devoted their lives to furthering scientific understanding.
As a result of their efforts, we now know more about ourselves
and our place in the universe than we have ever known before.

However, there are still plenty of questions left unanswered.
Right now, there are scientists working to unravel the infinite mysteries
of existence, including the details of the story that I have just told.
Some of the details that I glossed over in this book are up for debate,
while others are completely unknown.

For example, there's *dark* energy and matter:
we don't know anything about them,
except that we think they make up about 95% of the Universe!

In addition to the lack of certainty in the scientific community
regarding the details of creation, there are the creative liberties
I took for the sake of story, rhyme, rhythm and illustration.

If you would like to learn more about this story,
what it means and the science behind it,
please visit my website at:

www.JLDunbar.com

or make a trip to your local library.

If your curiosity still isn't satisfied,
maybe you should become an astrophysicist!

About the Author

James "Jamie" Lu Dunbar has written and illustrated two other books:
7 River Riddles and Gordy McGranite Grapples with Gradients: A Calculus Story.
Jamie Dunbar lives in Oakland CA, where he works for DogStarDaily.com,
the most comprehensive dog training resource on the internet.
Jamie went to Brown University where he majored in Sociology and Visual Art,
specializing in oil painting and bookmaking. In addition to art,
Jamie enjoys cooking, gardening, playing games of all sorts and of course...
...science.

Learn more at www.JLDunbar.com

Colophon

I wrote this book while living in Boston in 2006 & 2007.
I made regular use of the public libraries,
researching illustrated children's books,
graphic novels and the creation of the universe.
I began illustrating the book in earnest in 2009.
I did preliminary drawings in pencil and ink
to design the characters and the basic paneling.
I scanned these sketches and used them
as the foundation for my final illustrations,
which I made in Photoshop with a
Wacom drawing tablet and a MacBook.
I made the book using InDesign
and had it printed by CreateSpace,
an on-demand publishing service owned by Amazon.

This font is Gill Sans, the story font is Chalkboard.